ULTIMATE SUPERCARS

MASERATI
GRANTURISMO

By John Perritano

Kaleidoscope
Minneapolis, MN

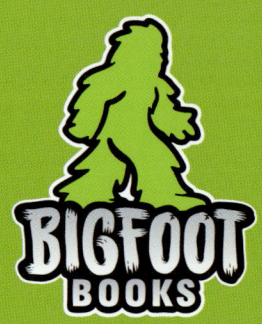

The Quest for Discovery Never Ends

This edition first published in 2021 by Kaleidoscope Publishing, Inc.

No part of this publication may be reproduced in whole or in part without written permission of the publisher.

For information regarding permission, write to
Kaleidoscope Publishing, Inc.
6012 Blue Circle Drive
Minnetonka, MN 55343

Library of Congress Control Number
2020936065

ISBN
978-1-64519-265-7 (library bound)
978-1-64519-333-3 (ebook)

Text copyright © 2021 by Kaleidoscope Publishing, Inc. All-Star Sports, Bigfoot Books, and associated logos are trademarks and/or registered trademarks of Kaleidoscope Publishing, Inc.

Printed in the United States of America.

FIND ME IF YOU CAN!

Bigfoot lurks within one of the images in this book. It's up to you to find him!

TABLE OF CONTENTS

Chapter 1: Arrivederci GranTurismo 4

Chapter 2: The Need For Speed 10

Chapter 3: What a Show! 16

Chapter 4: The Love Affair Continues 22

Beyond the Book *28*
Research Ninja *29*
Further Resources *30*
Glossary *31*
Index *32*
Photo Credits *32*
About the Author *32*

Chapter 1
Arrivederci GranTurismo

Connor Golden first saw the Maserati GranTurismo (GT) in 2007. He was in Texas taking pictures of high-end sports cars. The GT was one of them.

Connor was struck by the car's beauty. He was bug-eyed by its performance. "The car was in a class of its own," he wrote on his website.

The GranTurismo had competitors. The Aston Martin Vantage. The Mercedes CL550. The GT out-classed them all. It had style. It had flair. It had speed and power.

After all, it was a Maserati!

For 12 years the GT thrilled the luxury sports car market. But all good things must come to an end. The same is true for the GranTurismo.

THE TRIDENT

Maserati has a famous badge. The trident was a weapon of mythology. It has three points. Maserati has three goals with all its cars. They each should provide a great driving experience. They use amazing technology. And they are packed with luxury. Mario Maserati, a brother of the founders, designed the logo in 1920.

PARTS OF A
GRANTURISMO

Galvanized steel body

Golden traveled to the Maserati factory in Modena, Italy in 2019. He wanted to say *Arrivederci*. That means "goodbye" in Italian. Maserati was stopping production of the GT. The company still planned on making the car. However, the new GTs would be electric.

The last GTs to be made were in parts on the workshop floor. Those parts would be put

Antenna built into rear window

Forged aluminum wheels

together by experts. Nuts and bolts were lined up. Engine parts were in a pile. Maserati **badges** lay next to window glass. Tires were stacked in the corners.

Golden walked around. He looked at the Maserati parts. He knew the end was near. It was the end of an era. The GT would be no more.

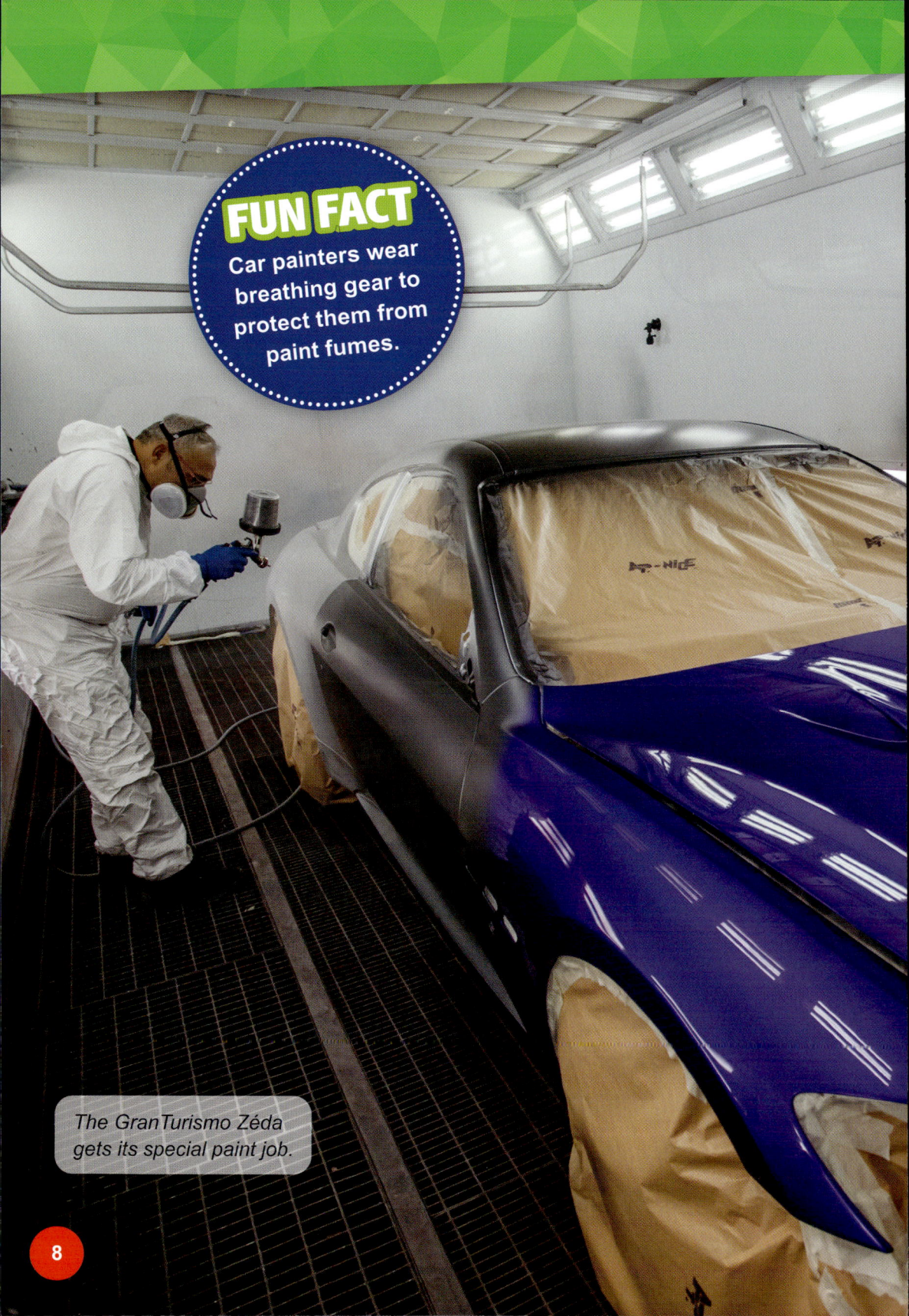

FUN FACT
Car painters wear breathing gear to protect them from paint fumes.

The GranTurismo Zéda gets its special paint job.

Maserati knew how important the GT was to its fans. It knew the GT was a winner. So it made the last model a very special one. The final Maserati GT was called the GranTurismo Zéda. That means GT-Z in Italian. That makes sense, since Z comes last, too!

Workers painted the car three colors. Blue, black, and white. Each color folded into the other. It was a work of art.

Golden still longed for the old GT. He thought it was one of the best car designs in history. The first GTs came out in 2007. The last GTs had the same style. The car ended as beautifully as it began. The GT was a tribute to Italian car making.

Chapter 2
The Need For Speed

Since the earliest days of racing, Italian cars have been out in front. Some of the most famous cars in the world were designed in Italy.

Maserati has always been at the top of the speed pyramid. The company began in 1914. It was founded

This very early Maserati racer is on display in a museum.

by three Maserati brothers—Alfieri, Ettor, and Ernesto. At first, they didn't build cars. They made spark plugs for World War I airplanes.

Alfieri Maserati drove race cars for other carmakers. He helped his brothers build cars, too. In 1926, the Maseratis put out their first car. The Tipo 26. It could speed along at 154 miles (248 km) an hour.

World War I lasted from 1914 to 1918. It was fought among European countries as well as the United States.

Maserati soon became a major player in the racing world. Tazio Nuvolari won races for Maserati in 1933 and 1934. In 1937, the brothers sold the company. That's when Maserati really took off.

The new owners put a lot of money into the design and performance of their cars. American Wilbur Shaw drove the Maserati 8 CTF to victory in the 1939 and 1940 Indianapolis 500.

When World War II (1939–1945) ended, the company shifted gears again. Instead of just racing cars, it built cars for everyday use. Still, Maseratis were fast and sporty. Among these cars were the first "granturismos," or GTs.

Maserati was a force in early auto racing.

WHERE THE MASERATI GRANTURISMO IS MADE

Poland

Germany

France

Modena, Italy

Italy

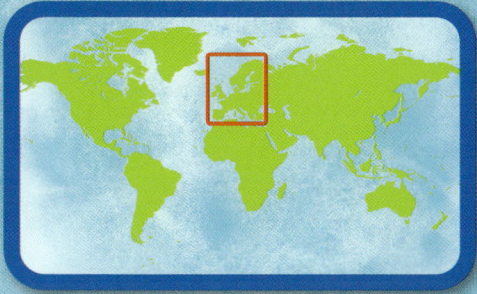

The first Maserati GTs came out in the 1940s. The original idea was simple. Put a race car engine into a luxurious **sedan**.

That's what Maserati did in 1947 with the A6 1500. It was the **prototype** of the first GranTurismo. It was the first Maserati people could drive on the road.

Designer Pinin Faria created a two-seat car. A long hood made the car even look fast! Its coolest feature was hidden headlights.

The public was stunned by GT's looks and speed. The GTs took many shapes over the next several decades. All were stylish. All were fast.

All were Maserati.

A rare Maserati A6 1500 that still runs.

MASERATI GRANTURISMO TIMELINE

1947: The A6 1500 is unveiled at the Geneva Motor Show.

1957: Maserati launches the 3500 GT Vignale Spyder.

1970: The Maserati Ghibli hits the road. It is named after a Saharan wind.

1998: Maserati builds the 3200 GT Assetto Corsa. It goes from 0 to 60 miles an hour in 5.1 seconds.

2007: Maserati launches its latest GranTurismo. It had a 405-horsepower engine.

2019: Maserati ends production of the GranTurismo. Its last model was the Zéda.

Chapter 3
What a Show!

Sergio Suppa enters the 2007 Geneva Motor Show. It is one of the largest car exhibits in the world. Suppa has eyes for only one car. He slowly walks around the 2007 Maserati GT.

Suppa is dressed in a dark suit. His black hair is slick and pushed back. He is an important man. Suppa runs one of Italy's largest banks. He also collects fast cars. He owns Lamborghinis, Porsches, and more. He circles the GT with his wife. She has long blonde hair and wears a sleek black dress. She carries a small purse.

Sergio looks at the GT. He blows it a kiss. "It can go from 0 to 60 in 5.2 seconds," he says.

"*Bellissima,*" his wife whispers. That means "very beautiful" in Italian.

FUN FACT
The first Geneva Motor Show was held in 1905!

The GT shines under the lights at a car show.

The brand-new GT made its debut in 2008. It was far different than the other models. It had Maserati muscle. It had Maserati quality. It has power and comfort. People can drive it to the store. They can tour along ocean roads. They can race it in the streets of Rome and Naples. Heads will turn, as the beautiful car glides by.

This new GT has room for four people. It doesn't lose one ounce of power. It can reach 177 miles (284 km) per hour. As Sergio said, it can go from 0 to 60 in 5.2 seconds.

The GT's **automatic transmission** is linked to a high-performance Ferrari 4.2-liter **V8** engine.

One review says its design is "to-die-for." Of course, it is. The designers at Maserati know what they are doing. It rivals Bentley's Continental GT and BMW 650i.

FUN FACT
Ferrari makes the Maserati engine. Many carmakers use each other's parts.

The front of the 2019 GT

The famous Maserati badge

THE 2019 MASERATI GT IN DETAIL

Height: 4 feet, 8 inches (1.38 m)

Width: 6 feet, 3 inches (1.9 m)

COST: $162,880 (United States)

LENGTH: 16 feet, 1 inch (4.9 m)

WEIGHT: 5,181 pounds (2,350 kg)

TOP SPEED: 180 miles per hour (289 kph)

TIME FROM 0-60 MPH: 5.2 seconds

"Look at the inside," Sergio says. The car has leather seats. The Maserati badge—the trident—is in the headrests.

The interior wood is polished. The car also comes with an 11-speaker audio system. It has a huge hard drive for storing music.

Maserati made the gas-engine GT through 2019. It was the most powerful **road-legal** machine in Maserati's collection.

Chapter 4
The Love Affair Continues

It's been 13 years since that Geneva show. Sergio's love affair with the GT continues. He loves the Zéda. He can't wait to get his hands on it. He travels to Maserati's factory in Modena. A company executive welcomes him with a handshake and a kiss on each cheek.

The factory has changed a lot since Sergio first visited in 2007. He went there after the Geneva Car show. Maserati changed its shop to build electric cars.

Maserati built 28,805 GTs in 12 years. Each one was built by hand. Sergio also wants to see what the new cars will look like. Three of Maserati's best models will be electric.

FUN FACT
Workers installed every part of the GranTurismo at the factory in Italy.

STEERING THE CAR

The GT has three steering wheel designs. They can be all leather. They can also be wrapped in leather and carbon fiber. The gear shift is long. It makes shifting easier.

"We have big plans for the GT," the executive tells Sergio. "But we are keeping the details secret for now."

For a while it looked like Maserati was going to stop GT production. Then, the company switched gears again. It decided to unveil a new electric-powered GT in 2021.

The electric GT's coolest feature is the lightweight carbon fiber hood. It also takes in air through snazzy side scoops.

The car is still designed to go fast. The GT has a front **spoiler** and side skirts. They keep the car stable at high speeds. Twin tailpipes help the engine work well. Of course, they also help it sound awesome!

"The sound of the engine makes your spine tingle, Sergio," the exec says.

"How easy will it be to get into the backseat?" Sergio asks.

The exec explains: both front seats are electronic. They slide forward and back. There's plenty of room in the rear seat.

The center arm rest pulls down. It has two cupholders and lights. Passengers can ride in comfort. There are climate-control vents.

Will the electric version of the GT be equally as powerful? Will it provide touring luxury with race-worthy performance?

Of course. And it will also help the planet.

Great design. Amazing technology. Luxury and style. That was and is the Maserati way.

FOR RACE LOVERS ONLY

The GranTurismo MC was the race-car version of the GT. It had a more powerful engine. A large rear spoiler helped keep the car on the track. They also added side skirts to make the car more **aerodynamic**. The company sold only a handful of these racers. The GT MC cost more than $150,000.

BEYOND
THE BOOK

After reading the book, it's time to think about what you learned. Try the following exercises to jumpstart your ideas.

RESEARCH

FIND OUT MORE. Where would you go to find out more about your favorite cars? Find out what company makes the car and locate its website. What information do the companies provide? What other sources of car information can you find?

CREATE

GET ARTISTIC. Cars start with creative artists and designers. Time for you to take a shot! Get art materials and create a great, new car. Will you make it a sports car? A sedan? A race car? What colors will you paint it? What features can you give it? Let your imagination go for a spin!

DISCOVER

DIG DEEPER. The GranTurismo is going electric. Look into how electric cars are different than gas-powered cars. How will they help the environment? What are the disadvantages of driving an electric car? Do you think more cars should become all-electric?

GROW

GO TO A CAR SHOW. Car shows are a great way to see lots of cool cars up-close. Check your local events calendar, or ask at a car dealer for upcoming events. You can find shows of old cars and new cars, sports cars and classic cars. Go to a show and find a new favorite car to love!

RESEARCH NINJA

Visit *www.ninjaresearcher.com/2657* to learn how to take your research skills and book report writing to the next level!

RESEARCH

DIGITAL LITERACY TOOLS

SEARCH LIKE A PRO
Learn about how to use search engines to find useful websites.

FACT OR FAKE?
Discover how you can tell a trusted website from an untrustworthy resource.

TEXT DETECTIVE
Explore how to zero in on the information you need most.

SHOW YOUR WORK
Research responsibly—learn how to cite sources.

WRITE

GET TO THE POINT
Learn how to express your main ideas.

PLAN OF ATTACK
Learn prewriting exercises and create an outline.

DOWNLOADABLE REPORT FORMS

Further Resources

BOOKS

Lumsden, Reece and Maks Zaretti. *The Maserati Life.* Seattle, WA: Amazon CreateSpace, 2019.

Monte, Dal Luca, et. al. *Maserati: A century of History.* Milan, Italy: Giorgio Nada Editore, 2014.

Oachs, Rose Emily. *Maserati GranTurismo (Car Crazy).* Mankato, MN: Bellwether Media, 2018.

WEBSITES

FACTSURFER

Factsurfer.com gives you a safe, fun way to find more information.

1. Go to www.factsurfer.com.

2. Enter "Maserati GranTurismo" into the search box and click 🔍

3. Select your book cover to see a list of related websites.

Glossary

aerodynamic: able to move through the air smoothly and easily.

automatic transmission: a system in a vehicle in which the car changes gears at different speeds without direct control by the driver.

badges: a metal disc or shape in the form of a car company's logo.

prototype: a sample car made to show what a production car can look like.

road-legal: having all the equipment in the car that allows a person to drive it legally on the road.

sedan: a car with a closed body with two rows for seating and a cargo trunk.

spoiler: an aerodynamic flap on the back of a car that is designed to "spoil" the movement of air across the body of a vehicle.

V8: A V8 engine has eight cylinders in the shape of a V.

Index

Aston Martin Vantage, 5
badge, 5, 7, 19, 21
electric car, 22, 25, 27
Faria, Pinin, 14
Geneva Motor Show, 15, 16, 17
Golden, Connor, 4, 6, 7, 9
Indianapolis 500, 12
interior, 21
Lamborghinis, 16
Maserati, Alfieri, 11
Maserati, Ernesto, 11
Maserati, Ettor, 11
Maserati 8 CTF, 12
Maserati A6 1500, 14, 15
Maserati GT MC, 27
Mercedes CL550, 5
Modena, Italy, 6, 13, 22
Porsches, 16
Shaw, Wilbur, 12
steering wheel, 24
Suppa, Sergio, 16, 18, 21, 22, 25, 26
Tazio, Nuvolari, 12
Tipo 26, 11
World War I, 11
World War II, 12
Zéda, 8, 9, 15, 22

PHOTO CREDITS

The images in this book are reproduced through the courtesy of: Courtesy Maserati: 4, 6, 9, 12, 18, 20, 23, 24, 25, 26. Shutterstock: Ivan Cholakov 11; Ying Gent 17; auto-data-net 21. Wikimedia: Supermat 14.
Cover: Santi Rodriguez/Shutterstock (car); fotomak/Shutterstock (background, top); zhao jiankang/Shutterstock (background, bottom).

About the Author

John Perritano is an award-wining journalist, author, and editor from Southbury, Connecticut. He has authored numerous books and articles on subjects such as science, technology, history, and current events. He holds a master's degree in American History from Western Connecticut State University.